The Future of Waves!

Deon Pollett

© *Copyright 2018 Light-Matters-Sanctuary.org. All rights reserved.*

No part of this book may be reproduced, stored in a retrieval system, or transmitted by any means without the written permission of the author.

Forward

Deon Pollett is a person that has tried it all. His many vocations have inspired him to write several books in the last year. Deon has a bachelor's degree in Business; A Masters Equivalency in Computer Science and a PHD in Metaphysics. He currently calls Northern California his home.

Some of his other books include

To Life – A book of prose and poetry
The Way – A search for truth in story form
The Lonely Eagle Speaks – A story of nature

This is a story about life that brings together ideas about how a company should be organized and weaves interactions with people on a planet with three moons and one sun called Trilla.

The author would like to thank Linda Schreiber for her work in editing and making suggestions.

Table of Contents

CHAPTER ONE 1
  'Values' 1
CHAPTER TWO 19
  'Metaphysics' 19
CHAPTER THREE 27
  'A business venture' 27
CHAPTER FOUR 35
  'Putting it all together' 35
CHAPTER FIVE 40
  'Visitors from space' 40
CHAPTER SIX 54
  'What to do, what to do?" 54
CHAPTER SEVEN 64
  'The Party' 64
CHAPTER EIGHT 79
  'Robots for the mines' 79
CHAPTER NINE 83
  'The Company You Keep' 83
CHAPTER TEN 94
  'Waves of the Future' 94
CHAPTER ELEVEN 104
  'Propinquity' 104

CHAPTER ONE
'Values'

Bob Truehorn had been working eleven hours this Saturday and he decided to go home. It was 6 PM and no one was around. He turned his computer off, and turned out the lights as he walked out into the hallway of the thirteenth floor and pressed the elevator button.

"I still can't quite get that program to do what I want it to do," he thought to himself. "Maybe after I rest a bit."

He left the elevator and went out on the main floor. No one was around except the guard who was sitting by the entrance to Lee Enterprises. He said, "Goodnight Toni"

"Goodnight Mr. Truehorn. Have a nice evening."

"I will," he said. "How late is your shift?"

"I won't be going home until midnight. I got here at 2 today. It's nice it's still light out there."

"Yes, summer is great. See you on Monday."

"Yes sir." said the Guard named Toni and Bob left the building.

Lee Enterprises owned the thirteenth story building and Bob had worked for John Lee for the last three years. Bob and John Lee were the only computer programmers in the company. It wasn't unusual for him to work sixty to seventy hours a week. This week was no different.

He walked down the bright street and noticed there weren't as many people on the street as usual. "Oh, it's Saturday," he thought to himself. "Most people only work Monday through Friday here in the city."

Bob lived in Riverton. Riverton was on the South Coast and the population around ten thousand people. Riverton was the largest city in the County of Billows. Billows county was one county in the State of Heatherland. Riverton was the county seat. The town was next to the Sea of Chibola and behind the town was a large mountain they called Mount Fanna. Mount Fanna was an old extinct volcano and there was a lake in the top of it. He liked to spend his Sundays on Lake Fanna fishing.

Tomorrow would not be different. He walked the two blocks to his apartment and unlocked the door. His bags were packed for the overnight trip to Lake Fanna so he picked them up and grabbed his fishing pole and fishing box and headed to the parking lot where his Porsche was parked.

As he drove up the mountain the sun started to set over the ocean and he decided to pull over and watch it go down into the water. There was a large overlook halfway up Mount Fanna and several people were parked and doing exactly what he wanted to do. Watch the sunset. The sky was a beautiful blue color and as the sun dropped into the water the sky turned a brilliant pink. At the exact moment the Sun could no longer be seen on the horizon, Bob saw a flash of green and it was gone.

"That's always worth watching," thought Bob to himself and he got into his Porsche and headed up the road.

Bob Truehorn was twenty two years old when he graduated from Leeder College in Computer Science some three years ago. The position at Lee Enterprises was his first good paying job out of college. Lee Enterprises sells teller machines to banks. Bank tellers have their Lee Enterprises computers that help them do the calculations and interconnect to the main computers at Bank's headquarters. He spent many of his days installing the software and helping the small company deliver the machines to Banks all over the Southern Coast of Heatherland. He also trained the new users on how to use the computers.

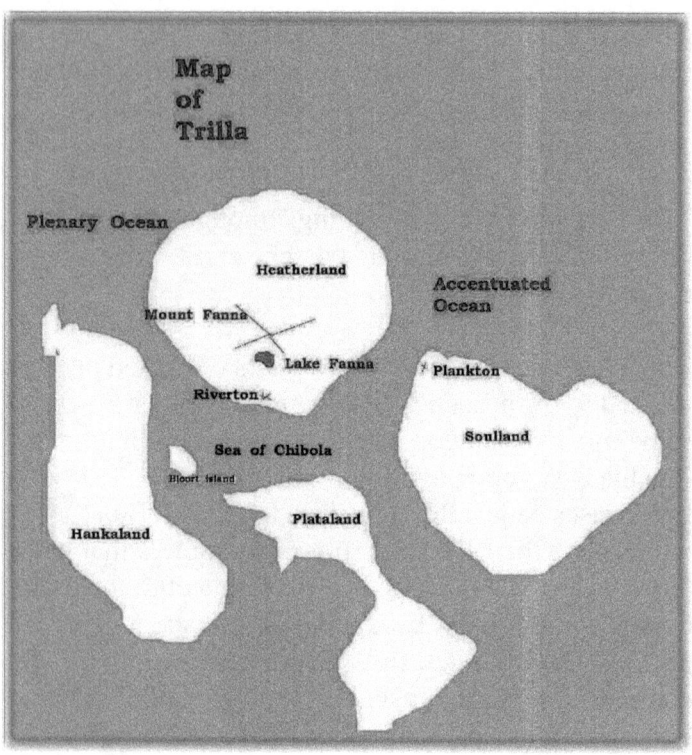

There were four large land masses on this planet. Heatherland was one of the land masses. The Ocean to the West was called the Plenary Ocean. To the East was the Accentuated Ocean. The water to the south they called the Chibola Sea. The other three land masses on Trilla were Hankaland, Soulland, and Plataland. Lee Enterprises was only on Heatherland. The owner John wanted to build up the company and put teller stations in the banks of all the four countries. All the banks in Heatherland bought Lee Enterprises computers, but it wasn't enough for John Lee. John Lee dreamed that his company would be all over Trilla in all the countries.

"It's good everyone speaks Trillish," thought Bob as he pulled his Porsche into the driveway of the cabin he had purchased on Lake Fanna. John Lee was talking about making Soulland the next country for them to install computers in. He was thinking of visiting the country called Soulland because they usually spoke a different language. Soulland normally spoke Soullish but most everyone there knew Trillish as well. Bob had traveled to Soulland and all the other countries in the world with John Lee the first summer he worked for him. John Lee had big hopes for his company and wanted to treat his employee's right. Bob went to work for him when he offered to take him around the world as part of the fringe benefits of his job. The Porsche was also a fringe benefit. It was one of the most expensive cars on Trilla. Bob had earned the Porsche and the world trip since going to work for John Lee some three years ago. He generally worked twelve to fourteen hours a day and most of the time he worked six days a week. It was nice to have Sunday off. That night he watched the three moons of Trilla come up almost simultaneously on different sides of the sky.

"What a beautiful sight," he thought.

The next morning John awoke with the Sun and admired the view of the pines and the lake out the window of the cabin he had purchased a year ago. He fixed himself a couple of eggs and a strip of bacon for his breakfast and took his fishing pole and fishing gear to the boat that was tied to the pier next to his cabin. It was a beautiful morning and he could hear birds singing as he put his pole and gear in the boat and untied it.

He started the little motor and went around the lake for awhile before he picked the spot where he would fish. Near the other side of the lake a spring bubbled up under the water and he could see the fish swimming around under his boat. It wasn't long and he had a half dozen fish strung on a stringer and was heading back to his cabin.

"This is always so relaxing," he thought to himself. "Someday maybe I'll find someone to share it with."

He had dated several girls in his short life but none of them had the things he had hoped for in a lifetime companion.

He unloaded his fish and cleaned them in the sink of the cabin and put them in the refrigerator. Then he set down at the table and pulled his laptop out of its cover and started looking at the program that wasn't doing what he wanted it to do. Within a few minutes he found the problem. He smiled to himself, and thought, "I should go fishing every day. That's all I needed to do was relax a little to find the problem."

Sunday went by quickly and he found himself in the Monday Morning meeting at Lee enterprises once again. John and six employees were setting around the table when he walked into the meeting room.

"Hey Bob," called John. "I'd like you to meet our new employee Hanta Khan. He is from Soulland and speaks the language. He is our new Marketing director."

Hanta was almost as tall as Bob. Bob stood about six feet and four inches tall and John Lee was only Five feet and three inches tall. Hanta was a mild mannered looking man and had a smile that filled the room. His hair was almost black.

"Very pleased to meet you," Hanta said as he extended his hand for a handshake.

"Likewise," said Bob.
The other five employees were Bill Grateful, the shop manager; Suzy Lundquist the Human relations manager; Smern Golden the quality control manager; Lacy Smith the research design manager and Dana Bloort the transportation director.

John Lee introduced Hanta around and said,

"The exciting thing to me is Hanta knows all the banks in Soulland. His past position is Sales Manager for the company that makes the bank vaults. He is going to help our company grow into Soulland."

"That is exciting," said Bill Grateful. "I hope we can produce our computers fast enough to keep up with your sales."

"Me too," said Hanta and everyone smiled.

Bill Grateful was a slender person with premature gray hair. He was in his thirties and married with one child. He was an excellent shop manager and was always able to meet the demands the sales department made on deliveries of computers and programs.

Suzy Lundquist spoke up and said, "I'll see to it that your paperwork is ready to go before the day is out. We do want to get you paid on time."

"That would be nice," said Hanta.

Suzy was a blue eyed blond with a slight problem of being overweight. She said it was because she liked to cook. Her job in Human Relations fit her as she loved people.

"Every one of our computers will meet the needs of our customers as fast as you can sell them," said Smern Golden the Quality control manager.

"Good to hear that," said Hanta and Bob and John Lee agreed.

Smern was a red head from Hankaland. He moved to Heatherland to work for John Lee enterprises. He was a good friend of Bill and they golfed and fished together. He was not married.

"Just don't go promising anything we can't deliver," said Lacy Smith the Research director.

"Don't worry. I won't," said Hanta.

Lacy Smith was a tall brunette. She wore horn rimmed glasses and was very intellectual. Bob thought of her as a teacher.

"We'll get them delivered faster than you can sell them," said Dana Bloort the Transportation Director.

"I'm sure you will," said John Lee taking control back of the meeting.

Dana Bloort was about six foot six and had dark brown hair and a large moustache. He bragged that he could press his weight and Bob believed him.

After the meeting everyone went to their respective places of employment and got busy doing their thing.

John Lee, the boss, entered the office where Bob was working and said, "How are you doing on that program you've been working on?"

"I've got it finished," he said and smiled. "You know, it's funny but I was having problems with it until I went to Lake Fanna and caught some fish yesterday. After that I figured the problem out easily."

"I knew you could do it," said John. "Look. We have orders already from Soulland for thirty computers and they want them delivered next week.

Can you go with me for a week to Soulland and help me install them and train the operators?"

"Sure," said Bob. "When do we leave?"

"The computers will be ready Friday night. I'm having them shipped directly to Soulland Bank of commerce on the North Coast to the city of Plankton. That's where the headquarters are for the Bank of Soulland. We have to be there Monday morning to install them and connect them to their main computer. It will probably take us most of the week so plan on being there for a week."

"Sounds good to me," said Bob as he returned to the next program he was working on.

The week went by quickly and Bob found himself flying to Soulland the following Sunday evening. His reservations were for a hotel in the city of Plankton. He visited Plankton once before but it had been two and a half years. As they landed he noticed things had not changed much. Soulland was a very flat country compared to Heatherland. Plankton was one of about fifty cities. The population in Plankton was close to one million people. A few rolling hills could be seen in the distance as the plane landed.

After they had checked into the hotel they had dinner and met at the bar. John had picked up a couple of beautiful women and was entertaining them when Bob arrived at the bar.

"Here he is girls," said John as Bob walked up to them. "The most wanted single man in Trilla. Bob Truehorn meet Tammy and JoAnne."
Tammy and JoAnne stood about five foot two inches tall and their bodies looked like they could be fashion models. They both had dark hair and their skin was a lovely tan color as well. Tammy had a turned up little nose and JoAnne had a smile that seemed to light up the room.

Tammy and JoAnne were very beautiful and Bob couldn't help wonder how little John Lee had found them so easily. It didn't take him long, however to find out how the little man had found them.

"Tammy is the president of the Bank of Soulland," said John. "And JoAnne is the Human relations manager."

"Very pleased to meet you," said Bob as he joined the threesome.

The evening went by quickly and Bob noticed John was being especially good to Tammy. As the night went on it was obvious John was going to go places with Tammy that night. Bob looked at the beautiful little girl next to him who called herself JoAnne and thought, "Naw! She's not your type." He tried several times to get in a conversation but all she wanted to talk about was her modeling and how she loved working with the bank in human relations.

The next day John and Bob met in the offices of the bank and John said to Bob, "Boy. Tammy is a handful. We banged all night long so it's gonna be up to you this morning."

"But what of your wife and two kids back in Heatherland," Bob said.

"Be quiet. These people don't need to know. Shhh! Mum's the word."

The respect for John Lee that Bob had, dropped down a little as John said, "You're too serious. Come on, I'll introduce you to the people you have to train this morning."

The week went by quickly but Bob had lost respect for his manager and boss since John and Tammy were openly doing things together all week.

On the plane trip home John said, "It's business Bob. I will do anything to make this company a success."

Bob nodded his head and said, "Well, I hope your wife doesn't find out or you'll be in deep trouble."

"She won't find out, unless you tell her," John said. "You won't, will you."

"I won't tell her," said Bob. "But I probably should."

"Again, you're too serious man. Lighten up."

John and Bob arrived back at home and now somehow Bob didn't feel like working as hard to make Lee Enterprises a success.

Six months went by and they were busy flying back and forth to Soulland installing computers and training employees. Bob became very good friends with Hanta Khan. Hanta had a home in Plankton and he was renting a condo in Heatherland. The second week Bob was in Soulland Hanta invited him over to his home for dinner. Hanta had a beautiful home right on the beach. Hanta's wife was a beautiful lady. When Bob arrived for dinner, Hanta introduced his wife, "Georgina, this is Bob. He is our top programmer and trainer for the company."

"Hi Bob," she said. "I hope you like spaghetti."

"I love it," said Bob.

"You and I are about the same age," said Hanta at dinner. "You're about twenty five, right?"

"You got it right on the nail," said Bob.

"So how come you're not married yet?"

"The right girl hasn't come along," said Bob.

"Maybe I could introduce you to some of my friends," said Georgina.

Georgina had hair as black as Hanta's and her skin was a golden color. She was almost as tall as Hanta. Bob guessed she was about six foot, two inches tall.

"Hey. I don't need any help finding girls, but thanks for thinking about me."

Dinner was wonderful and Bob had found friends for life, he thought.

The next six months the crew installed computers at almost every bank in Soulland. Bob even learned to speak the language.

One Monday morning after the meeting in Heatherland John Lee asked Bob to come to his office.

"Bob, I want to tell you of my plans because it's important you go along with me," John said. "Our company has grown immensely since Soulland and now the banks of Hankaland and Plataland want our computers. The problem is we are hurting for capital. We need money. I have been talking with a possible benefactor in Plataland in the city of Greenoble and he has given me an offer. Before I take it I wanted to talk to you, because you really helped me start this company."

"So, what's the problem," Bob asked.

"Well this investor has just the amount we need. The only problem is he wants to be the Marketing

Manager if I take his offer. I haven't talked to Hanta about it, because we can't have two Marketing Managers."

"It seems to me, the answer should be 'Sorry, but I'm not interested.'" Said Bob. "Hanta has been the reason this company is doing so well. Besides I like him."

"I know you feel that way. That's why I wanted to talk to you first. If I say yes to this man and fire Hanta we could have this business ready to sell within six months and give us all enough money to live on for the rest of our lives."

"It all sounds too good to be true." said Bob. "I don't believe in treating your employees the way you would treat Hanta. You've used him and now you want to discard him like some old worn out blanket."

"Listen. I've pretty much decided to do this. I just wanted you to be with me. You've helped more than anyone can say with this company. I'll make you a full partner or you can have any position you want with the company."

Bob looked at him and said, "You can't be serious. If you are then I don't want to have anything more to do with this company. I'm quitting."

John looked sad and said, "I'll give you anything you want. Any position. Please go along with me."

"Sorry John," said Bob. "I was looking for work when you hired me and I can do it again."

With that Bob got up and left John's office. After about two hours, he had packed up his desk, so he told Hanta, Bill, Suzy, Smern, Lacy and Dana goodbye. Bill pulled him aside and said, "Confidentially, man; you're making a mistake. We're gonna be rich."

"Maybe so," said Bob. "But I don't want to be any part of it, the way John is treating Hanta."

He left the building and decided to go fishing before he started looking for another job.

Hanta called him on the phone and said, "When you resigned this morning, I had no idea it had anything to do with me, until John fired me this afternoon. What are you going to do?"

"I don't know, but do you want to go fishing with me at Fanta Lake? I have a cabin up there. You can bring Georgina if you'd like. It's a three bedroom cabin so you and Georgina could have a room to yourself."

"Maybe that would be a good way for me to tell Georgina," said Hanta. "I'm unemployed. Thanks for the offer. When are you going up?"

"Right now," said Bob. "Get Georgina and come on up. I have plenty of food and there is lots of fishing."

"Sounds great. I'll have Georgina fly our little plane over and we'll see you in a while. My plane can land on water, so we can fly to the lake." Hanta hung up the phone and called his wife and invited her to come to Heatherland.

That night Bob, Georgina and Hanta celebrated together and said as they were toasting, "To new beginnings."

Six months later, Lee Enterprises sold their company to International Trust Corporation. Bill Grateful called Bob to gloat.

"Hey Bob. We just sold the company and John gave all of us part of the fifteen million he sold it for. You could be a millionaire if you had stayed with us."

"Thanks Bill," said Bob. "You make me real glad I stood up to my values and didn't support John's treatment of Hanta. What's happening with everyone else?"

"Well, John Lee is getting a divorce. Somehow his wife found out about his playing around. She is taking him for most of the money he made in this sale. His two kids will live good. Meanwhile those of us that ignored his actions and went along are retiring. Suzy moved to Soulland; Smern bought a big ranch in Hankaland; Lacy is a millionaire but still working for International Trust Company. She is in charge of things in Plataland. She told me she had to work to survive; and Dana bought an island off the coast of Hankaland. And me, why I just bought a ranch on the north side of Heatherland. Four hundred acres and cows. What's happening with you?"

"I'm still unemployed, but I've been doing odd programming contracts for a few small businesses. Enough to keep me alive, anyway," said Bob.

Bob went to sleep that night thinking he could have been a millionaire, if he had only NOT listened to his conscience. "I have my values still," he thought to himself and he put the tackle in his boat and went fishing.

CHAPTER TWO
Metaphysics

Bob started playing mandolin every Wednesday night with a blind guy named Ralph Burgess. Ralph played guitar and sang and Bob played his mandolin in the background. It was fun for Bob and Ralph said he really liked the mandolin in the background.

One evening at the bar where they were playing, while Bob was on break a man came up to Bob and said, "I don't like your looks."

Bob looked at him and thought, "Remember, don't take anything personal."

He said out loud, "What are you going to do about it?"

"I'm going to clean your clock."

Bob stood back and took assumed a defensive stand he had learned in Aikido as the man rushed him. Before the man reached Bob, two doormen/bouncers jumped the man and yelled, "No fighting!"

Bob stood there aghast as the man proceeded to get one of the doormen in a headlock and remove one of his eyes from its socket. The doormen's eye was hanging listlessly at his side. The other bouncer subdued the man and the bartender called the police and an ambulance and hauled the person that started the fight away and took the bouncer that was hurt to the hospital.

Bob watched the entire incident in disbelief. He couldn't respond but was ready to. His method was to be defensive, not aggressive.

Bob continued finding odd jobs in Riverton until one day he decided to go back to school and get his Masters degree. He applied at the school and decided to start that fall.

At school Bob decided to study psychology instead of math and computer science. It seemed it would be more fun to him to learn about the human mind. The course at Leeder College was called the Leadership Institute of Billows or LIOB. It was highly recommended by a cute little psychic girl he had met and liked named Zaza Plinkton. She was working as a marriage counselor and only a few people knew of her psychic abilities.

Bob joined a group called the Psychic Explorers. They would meet at the college once a week and discuss paranormal events and abilities. He found he didn't have any paranormal abilities, but it was fun to watch others channel people from the other side; tell of visiting the Akashic records (which are records telling you of all your different lifetimes and why you are here on the earth this time.)

Bob wasn't sure he believed in re-incarnation but it was interesting to study along with his psychology studies at the Institute.

One evening at the meetings, a Buddhist priestess attended. She looked to be in her seventies but Bob was impressed by her. After the meeting he asked if he could study Buddhism with her.

"Well, I don't just take anyone," she said. "But you look like you're serious so I'll give you six months lessons for free. Come to see me every week on Wednesday's at 4 PM. Will that work for you?"

"That will be neat," said Bob.

The following Wednesday he went to the Priestesses home to study Buddhism. The first thing she told him to do was to repeat the chant 'Om, padmay seedee hum' over and over again for at least ten minutes every morning. She told him this was the only chant he would ever need.

During the next six months Bob would start out every morning chanting and sitting in meditation.

On about the fourth month the priestess told the psychic explorer group she would teach them how to see their past lives. This would help them to know why they are here and their purpose on the planet Trilla in this lifetime.

She turned out the lights in the room and lit a couple of candles. Then she had a couple of people help her set up a mirror. One by one she asked us to come and sit in a chair next to the mirror and look at yourself in the mirror. The light was low but it was easy to see yourself and the class around you. She told us to stare at only ourselves in the mirror. What was supposed to happen was your face would start to change.

"Do not be alarmed," she said. "Allow the faces to change and keep watching. Usually within a couple minutes you will see several faces. They will flash over and over again and gradually you will see one face a little longer. The longer you look into the mirror the more you will learn about each of the faces. These are your past incarnations and you can learn as much as you want about each one just by being calm and looking at each face."

When it became Bob's turn, he was excited. Soon he saw several faces flash before him. As he watched the changing faces slowed down and a little more time was spent on each one.

Bob saw Himself as a member of the mafia and a drug lord. It scared him a little but he kept watching and he saw a Sumo wrestler in Plataland. Then he saw what looked like a sheepherder in ancient Heatherland. "Enough," he thought and the faces disappeared and Bob could only see his image.

He told the group he saw himself as a member of the mafia and it scared him. The Priestess said, "Do not be afraid. You can do this at home or anywhere there is a mirror with low light. These images will show you what you are to learn in this lifetime."

At the end of the meeting he asked Zaza for a date. She said, "Okay, but I'm really not interested in anything like a relationship."

"You do have to eat, don't you?" asked Bob.

"Of course, we all do," said ZaZa.

"All I'm asking is for your company at a nice place for dinner, I'll even pay," said Bob.

"I wouldn't hear of it," said ZaZa. "We'll go dutch. Each of us will pay for our own meal. If you want to do that I'll go with you. Have you eaten tonight?"

"No," said Bob surprised. "Would you like to go somewhere and eat?"

"I'd love to," she cooed and batted her eyes at him.

"For a girl that doesn't want a relationship, You sure are acting like you want one," he thought to himself.

"I heard that," she said. "Just because I flirt a little doesn't mean I want to have something serious between us."

Bob was taken aback by the fact she could here his thoughts and decided to talk about something she could tell him about while they went to a restaurant next to the college and ordered dinner.

"Tell me about the Akashic records," he asked as they ordered their meals.

"There is not much to tell," Zaza said. "I have been able to visit them and look up records for people since I received these abilities that I have. They are records about your soul going back in time here on Trilla and on other planets if you've lived on them."

"Okay then. Tell me about how you started doing psychic things."

"It all started one day when I was driving home from work as a counselor. I was listening to the National Public Radio on my Porsche and thinking about how awful the news was. I pulled my Porsche off the highway and prayed, Please God, help me do something to make this world a better place. I prayed with much emotion and felt like I was at the end of my rope listening to all the problems of the world. That night I saw a vision of an old lady. She came out of the walls and blood was streaming down the wall. Then she came into my body. Since then I have been able to see auras, visit the Akashic records and do all manner of things psychically."

"Wow," said Bob. "Can you visit the Akashic records and tell me something about one of my past lives?"

"Sure. Hang on a minute."

She closed her eyes and said "I see one of your lives that is still affecting what you do. You were a sheepherder in that life wandering around in the desert herding sheep. You were very sad because you were in line for the throne to be the new king of the country when your father died. He had promised you the kingdom but when he got old, his memory failed him and he made one of the other people in the court the king before he died. You took it pretty hard and became a sheepherder. You wandered in the desert herding your sheep for the rest of that life. You still carry the wound in your heart. It affects your trust of people and especially people in charge. Have you ever had problems with your bosses?"

Bob told her the story of International Trust Corporation. She said, "The world will keep dealing different things to you until you get over the hurt in your heart. It is probably what is keeping you from having any psychic abilities."

They finished eating and went to their respective homes alone. Bob thought, "I have received a bunch to think about today" and then he went to sleep.

One year later he quit going to college as he was pretty much out of money. He started looking around for work and decided to try something he had never tried before.

CHAPTER THREE
'A business venture'

During the next week or so, Bob decided he'd like to start a business of his own. He played guitar and knew a little about the fiddle so he decided to open a music store that catered to bluegrass musicians. He found a source for the less expensive guitars, mandolins and fiddles and took the little money he had saved and invested it in leasing a building.

He hired a woman to help him behind the counter and to help with the bookkeeping. Her name was Ruth Zablinski. She stood about six feet tall, which is tall for a female in Heatherland, and she had long black hair and brown eyes. Bob liked her very much, but he had read somewhere it was not good to date people you worked with, so he was friendly to her and appreciated her help in ordering music books, and the musical instruments the store needed to satisfy customer needs. He found out over the next few months that she was also a good sales person.

Many of the customers asked for Manton guitars. The Manton guitar company had a restrictive method of giving out territories for their sales force and Bob could not get the franchise to carry the Manton guitar. An existing music store in Riverton already carried the Manton guitars so he could not have the franchise.

Some of his customers asked him to carry the electric Blender guitar and it was the same story. The representative from the company told him there was already a music store in Riverton that carried Blender guitars.

Bob found himself working very hard and long in the store. Even though he had Ruth helping him, it seemed like there were never enough hours in the day. One day two customers were talking to Ruth about which guitar was best, and Bob overheard them and listened in.

"Manton guitars have the better sound of any other guitar," said one of the men. "I was born in Soulland and almost all musicians there buy only Mantons."

The other gentlemen looked at him and said, "You probably speak Soullish, then."

A third man joined them in the conversation and said, "I've never learned Soullish, but I think the guitars here are just as good as the Manton and they are much less expensive."

"That's what I think too," said Ruth to the three men.

A fourth man joined in and said, "I've been trying these guitars out and they do sound almost as good as my Manton at home."

"Say something in Soullish," said Ruth. "I'd like to hear how it sounds."

Next the first two men started talking away in Soullish and Bob listened intently. The one man was describing the neck of the Manton guitar and likened the body and the head of the guitar to a beautiful woman. Bob thought to himself, "I'm glad I learned Soullish."

Ruth said, "What did he just say?"

The third man said "He just told you, you had a fish bone caught in your throat."

"Ha!" Ruth said and she laughed as if he was joking.

The fourth man said to the third man, "I didn't know you knew Soullish?"

"I don't," said the third man. The conversation then turned to how the fourth man thought he was speaking the truth about having a fish in Ruth's mouth.

Ruth said, "That's the difference in men and women. Men usually talk BS. I knew you were joking when you told me I had a fish bone caught in my throat. You" (she pointed to the fourth man) "thought he really knew Soullish."

"I guess I'm just naïve," said the fourth man. "I don't know Soullish. But I just assumed you understood when you acted like you understood." He pointed to the third man.

"No. Ruth is right. I'm full of BS. But I still think these guitars are a great deal for the money. I'd like to buy the one I've been playing."

Ruth put it in a case for him and took his money while Bob shook his head and went into the back room to work on one of the guitars that someone had brought in for repairs.

After about six months in operation, Bob was still not making enough money to pay himself. He had lost fifty pounds since starting the business and was looking over the budget and the sales book one evening and thinking about his adventure into the business world.

He was three months overdue on the lease of the building. He still owed money to several creditors who had sent him inventory and every month after paying Ruth he was almost broke. He almost had enough to buy himself a little food once in a while.

"This is not worth it," he thought to himself that evening as he was going over the books. "I'm going to have to close up the store and find work."

The next day he told Ruth not to come back and he hung a sign on the door 'Going Out Of Business. Everything must sell.'

He sold most of the guitars, mandolins and fiddles and sent some of the music books and movies back to the suppliers with a letter stating he was out of business.

He then went back to his little house on the lake. He had given up the home in town after only two months in business trying to make ends meet.

He fell asleep and rested for the next week. Finally he got up the energy to go fishing. It seemed to renew his spirits catching fish.

He decided to give Hanta a call and see what was happening with him. Hanta asked him about the business venture.

"I heard you opened a music store, of all things. How did it go?" said Hanta warmly.

"Terrible," said Bob. "I couldn't sell enough to make the rent payment, let alone enough to pay myself. I called to see if you and Georgina would like to come up and go fishing."

"We'd love to. Is next weekend a good time?" asked Hanta. "I've got a great job as Marketing director for Manton guitars. I just started with them last week or I probably could have helped you in the music business."

"That's the story of my life. Everything comes around a little late," said Bob. "I'll look forward to seeing you this weekend."

The weekend came before Bob could blink an eye and he enjoyed Georgina and Hanta's visit. On Sunday before they went back to the city Hanta said, "You know there is an opening for a salesman in Soulland at Manton. If you're interested, I'll fix you up."

"I don't know," said Bob. "I seem to be a buyer more than a seller. My short business experience taught me that. I was lucky to have Ruth around. I think she could sell ice to the Eskimos in the North. I'll tell her about the opening if you'd like. I think I would like to go to work for the Government. They have paid vacations and seem to be the most reliable kind of business around."

Hanta said, "Please, tell Ruth about the position. Tell her to call me."

"Will do," said Bob. After dinner and a wonderful weekend, Hanta and Georgina Khan left for their home in Plankton, Soulland. "Have a safe journey home."

Hanta and Georgina flew in their little four seat airplane back to Plankton.

The next day, in the late afternoon, Bob went to Ruth's house and rang the doorbell. When she came to the door, Ruth said, "Bob, it's good to see you. What brings you here? Come in and sit down."

Bob entered the home and looked around. It was a small nicely furnished flat in Riverton. When Bob told Ruth about the position with Manton, Ruth got very excited.

"But I don't know Soullish," she exclaimed.

"I'll give you a cram course," Bob said to her.

"That would be very nice," she said and blinked her eyes at him in a way he'd never seen before.

"I've just taken a job in Soulland," said Bob. "It seems their government needs computer programmers for their military. I start in two weeks. I think I've got just enough money to make it that long."

"Well, I know you paid me a wage when I worked in your store even though you couldn't afford it and I appreciate it," Ruth said batting her eyes again at him. "I do not require much to live on as my folks left me this flat when they passed away. I saved almost every penny you paid me. Stay here for dinner and we'll talk about how you can give me a crash course in Soullish."

"I'd like that," said Bob.

Bob stayed for dinner and they talked until about two in the morning.

"Gee, I didn't know we talked so long," said Bob. "I'd better get home."

"Okay, but remember you have to start teaching me Soullish," Ruth said again batting her eyes at him.

"The first word for you is Maloha. That word has many meanings in Soullish. It is used to say 'See You Later', Hello, and Goodbye. It is a word that almost means let the spirit be with you in wherever you go. So for tonight, I will say Maloha sweet Ruth. Maloha."

She blinked her eyes and closed them and said "Maloha Big Bob."

Bob left for home and for some reason he was feeling full of energy. He had a difficult time going to sleep even though it was about three in the morning when he reached his cabin on the Lake.

CHAPTER FOUR
'Putting it all together'

Ruth learned Soullish fairly fast and within the two weeks she found herself working for Manton selling musical instruments and things to music stores in Plankton.

Bob Truehorn started working for the Soulland government as a computer programmer. He found a nice little home to rent in Plankton and decided to give Ruth a call after his first few weeks at work.

"Ruth, I got your number from the phone company. They told me your new number so I guess you've moved. It's been a week since I've seen you. Are you living in Plankton?"

"Yes. I found a little house that I couldn't pass up. I bought the thing and quickly sold my place in Riverton. Georgina and Hanta had me over for dinner last week and I love the new job. Thanks for suggesting me to Hanta."

"Hey! That's what friends are for."

Ruth paused for a moment and then said, "How do you like your new job here in Plankton and where are you living?"

"Oh, I found a small place to rent. I don't know if I'll like working for the military. We're designing a system for space travel. I don't know much about space travel, but I'm learning."

"That sounds pretty technical."

"Yes, it is. The problem we're having is after we design the system it will probably get farmed out to private industry for programming. My boss told me the government was in the business of getting out of business. I love to write code, but it looks like those good jobs will be farmed out to private industry."

He paused for a second and then said, "The reason I'm calling you is I want to take you out to dinner. Would Friday evening or Saturday be best for you?"

He was thinking to himself, "Don't ask for a date, and give her options."

"Saturday would be better for me," she cooed over the phone. "What time do you have in mind?"

"How about six thirty. I found a neat little place that serves oysters on the half shell. Do you like that kind of food?"

"I love it," Ruth said. "Will you want to pick me up here?"

"Sure, I can do that. Give me your address and I'll see you at six thirty on Saturday."

"I live at 3334 Montrose Avenue. Do you know where that is?"

"No. I installed a GPS system in my Porsche and it will tell me where you live and how I can find you. I'm looking forward to seeing you again."

"Likewise," said Ruth and they hung up their phones.

That week went by quickly and Bob punched the address into his GPS system. As it turned out she was only two blocks away from his house.

They went to dinner and had a marvelous time. After dinner Bob invited Ruth to his place and they talked again until the wee hours of the morning.

"I'm not in a hurry to get into any kind of relationship," she said to Bob as the clock approached three a.m. "I'm going home now and I love to talk to you. Tomorrow we both have off, so we can sleep in. I'll see you later."

She stood and said, "Oh, I forgot. You're driving."

"No problem," Bob said as he smiled and looked at her longingly. "I'll take you home. It's only a couple of blocks."

They got in his Porsche and went to her house without saying much. When she got out of the car she said, "Thanks for a wonderful evening. Let's do it again some time."

"You got it. Maloha." said Bob and she walked into her house. Bob watched her go and then returned to his house and slept until nine A.M the next day.

"I think I'll go visit that church on Hadnell that people have been talking about. It's supposed to be pretty metaphysical," he thought to himself. He got up, showered and shaved and dressed for church.

When he got there he was surprised to see many of the people that had attended the psychic explorers at the college including Zaza.

After the service Bob cornered her and said, "How long have you been in Soulland? I thought your business was in Heatherland?"

She looked at him and smiled. "After you left the college and the group, many of us moved here because it's more liberal. I do twice the business here that I did in Heatherland as a marriage counselor. The people in Plankton are more our kind of people. Do you want to go to lunch with a bunch of us at the local café?"

"Sure," said Bob.

"We're eating and discussing the service at Suzy's Café over on Hadnell and Vixon street. Do you know where it is?"

"Yes. I'll see you there."

At the lunch Suzy Lundquist said, "Bob do you remember me? I was the Human Relations manager for Lee Enterprises on another world far, far away."

"Of course I remember you Suzy. Bill told me you are a millionaire now."

"Yes. Well having all the money you want isn't what it's cracked up to be," said Suzy. "It's good to see you. I'm running a catering business and restaurant here in Soulland. It's doing quite well. In fact I own this restaurant."

"I didn't make the connection Suzy," Bob said. "Suzy's Café. I should have guessed."

After the breakfast everyone hugged each other and went their separate ways. Bob thought to himself, "Now that was fun."

CHAPTER FIVE
'Visitors from space'

When he arrived at his apartment there was a message waiting for him on his phone. He played it back and it said he needed to report to work immediately even though it's Sunday.

Bob got in his Porsche and went to the military installation and into the office where he had been working. Guards were all over the place and several of his co-workers met him as he entered the office.

"We just received this communications from off-planet," said John Ticker his supervisor. "It says we must show them how we are using our military space research for good and not for bad."

"What!" Bob exclaimed. "Now some off-planet personnel are telling us what to do?"

"These creatures are from the planet Sarduk," explained John. "It's a planet that circles our Sun every thirty six hundred years. The people on that planet are giants compared to us. My intelligence says they are hell bent on taking over our planet and using us as slaves to mine other planets. We have to convince them we are peaceful. We also have to convince them we are sophisticated enough to defend ourselves if they try to take over."

"I'm just a programmer," Bob said. "What does this have to do with me?"

"You have been in charge of our flying machines programming. You need to convince them we have this technology."

Just then one of the guards approached them.

"Mr. Ticker, You'd better come quickly. Several large spaceships are landing on our landing strips and they are from Sarduk."

"Bob, get your crew together and set up a demonstration of our flying ship in hanger thirteen immediately. I'll talk to the emissaries from Sarduk and persuade them to watch our demonstration."

"Okay. I'll need at least thirty minutes to set it up in hanger thirteen. Stall them if you can."

With that they left each other. Bob went to hanger thirteen with his crew and John Ticker went to meet the Sarduks.

He had a television and high fidelity transmission set up at the hanger from it to the control center in his office. The plane was in hanger thirteen and it was not operational as of yet.

Three very large people accompanied John Ticker into the hanger as Bob and his crew was making the final connections. He stopped what he was doing when they entered. He had never seen people that stood about sixteen feet tall. John Ticker was six foot six and was dwarfed by the three people accompanying him.

"Bob, I'd like you to meet Truhann, Blitz and Bjorg from Sarduk. They bent down to shake his hand and smiled.

"Mr. Ticker tells us you have created a program to make the flying machine work. We'd like to see an example of your technology."

"Certainly," said Bob. He pushed the button and the screen hummed and went blank.

"Just as we thought," said Blitz the largest of the three Sarduk giants. "You haven't the technology and you are using inferior equipment to make things happen."

"No. No. You don't understand," exclaimed Bob. "I haven't had enough time to show you our expanded mathematical equations that you will need if you are going to mine and travel throughout the solar system.

"What are you talking about," said Blitz. "We can travel throughout the solar system now with no problems."

"Yes, but we have discovered in our research the equation for how to turn base metal into Gold and you really need to look at this equation and see the kinds of research we have been doing."

Now it was Truhanns turn to speak, "Blitz that is the reason we have not taken control of these Trillings. We want to find out what the equation is. Our research is close but we haven't found it. Mr.Truehorn, I assume you have found the equation?"

"It is really a series of equations and how they work together that show us how to change the molecular structure of any item. I have them in my control room and with the correct security codes I can download them here. They will change the isotronics in the Space Ship and you will see it will fly without any power. In fact, it will not be bothered by gravity."

"If this is true your planet may be saved from our taking it over and making you our slaves," said Blitz.

"Of course it's true," said John looking hopefully at Bob.

Bob punched in a few codes to the communicator and the screen started to show some equations. Then it went blank again. He tried to wipe off the screen with a cloth nearby and tried it again. It hummed and died.

The smallest of the three giants, Bjorg said, "We've had enough. It's obviously a ploy to make us believe you are more sophisticated than we thought. Call in our other space ships."

Blitz and Truhann turned away and seemed to be talking to others on their communicators. Soon the sky above the hanger was full of space ships.

Bob and his crew and John Ticker walked out of the hanger and watched in disbelief as a huge mother ship descended and at least fifty other ships filled the sky.

"Is this the end," thought Bob. Out loud he said, "You cannot make us your slaves. There are too many of us."

"We have our ways," said the Giant they called Bjorg. The giant smiled at Bob and then suddenly disappeared.

"Where did he go?" Bob asked. "This is too much."

"He is now aboard the mother ship," said Truhann. "We can transport people from one ship to another by locking on them and using our multiplexer machine. Your discovery of the equations to restructure the atoms in items is a similar thing to what we do, but we have lost the equations. If you would be kind enough to get them from your office we will leave you in peace."

"You will have to give me an hour or so. I want to talk with John and my crew before I give anything to you. Can you give us an hour?"

"Surely, Mr. Truehorn. You and Mr. Ticker take your time. We will leave you now and be back in one hour."

With that he and Blitz disappeared and the mother ship and fifty ships all left their vision.

"I wonder how they disappeared so fast," said one of the crew.

Bob said, "I don't know but we need to get together and make some plans and fast. They said they'd be back in one hour."

John Ticker, Bob Truehorn and the crew of six went to the conference room in the main office. Bob used his phone to call Zaza Plinkton.

"Zaza, help! Some giants from the planet Sarduk are claiming they will take over our planet if we don't help them and give them the equations I've been working on. They want to make us all slaves. Can you pick up anything psychically that will help us?"

"Give me a minute," she said over the phone and silence filled the room. Bob put the phone on speaker phone and waited.

"Okay, I've got a handle on three of the people from Sarduk. Their names are Blitz, Truhann and Bjorg. They seem to be the leaders and they believe your equations will make their science not work. It has something to do with undoing the matter in the wave forms that travel throughout the galaxy. I see them as afraid of you and what you can do to them. It feels like a bluff to me when they threaten you that they will take over. Let me check the Akashic records on these guys. Hang on."

Again all was silent and Bob was scratching his head. Soon Zaza came back on the phone and said, "Just as I thought. These guys were the little people in one of their past lives and now they are trying to make up for the inferiority they felt in that lifetime by browbeating and threatening. I don't think they are dangerous."

"They are pretty big. And they seem to have many helpers. I saw at least fifty ships in the sky with them."

"Well, fifty ships isn't enough to take over Trilla. It might get a good start on the city of Plankton, but if we can get the word out to other cities to defend themselves they will not be able to take over."

"How do you suggest we defend ourselves," Bob asked her?

"Start by being just as aggressive as they are. Tell them it's no deal and you will destroy their science if they try anything. Then if they try anything run your program on the space ship you guys have been working on and they will see it work."

"How did you know about that? It's been kept very secret.'

Zaza said, "I know a lot of things, Bob. Remember, show some aggression and don't be afraid."

She hung up and Bob looked around the room.

"Are you nuts?" said John. "Talking to a psychic. It sounds like she knows what she's talking about, but what if she doesn't?"

"What do we have to lose?" one of the crew named Rick, asked.

"Rick's right, John. What do we have to lose? I think we need to bring the ship over here instead of trying to send the equations and the program over the wires."

Rick spoke up again and said, "I've got a thumb disk we can load the programs on and then take them to the ship. I don't know why we didn't think of that in the first place."

"I guess we were not thinking too straight when we saw the ships in our skies," offered John Ticker.

"I guess not," said Rick.

Bob and the crew loaded the program onto the thumb drive and left for the hanger. On the way there the skies again filled with the planes from Sarduk. Now fifty people were between them and hanger thirteen. All fifty of the people were at least sixteen feet tall.

"You can't make your plane operational," said Bjorg. "We will not allow it."

"I found out the problem with our communication device and fixed it. You cannot stop us," said Bob matter-of-factly.

"We don't want to stop you. We just want to see the equations. You must show them to us."

Rick jumped around ten of the people and Bob jumped the other way. His crew went right down the middle toward the hanger and John ran backwards.

The fifty giants easily stopped the crew but Rick was inside the hanger. Bob spoke very loudly,

"You have come here for the equations that will help you mine and travel around faster than you have been. If I initiate the sequence your science will not work any longer. It's important you do not stop us or I will initiate the sequence. If you will give us something in writing about not taking over Trilla we will be glad to share the equations with you. Now release my crew or I'll initiate the sequence to power up our ship."

The giants looked at each other and Bjorg shouted, "He's bluffing. Don't listen to him."

Truhann spoke up and said, "NO! We have traveled here to find out the equations. Let them go and we will meet with Bob and his crew inside again."

"Whew! That was close," thought Bob. The giants followed him and John into the hanger again.

This time Bob felt like he was in charge.

"I need something in writing saying you will not take over Trilla if I give you these equations. I want it now!"

There was a mumble among the Sarduks. Bjorg said, "Don't do it. We don't need it."

"Okay, Rick. Activate the program."

Rick stood on top of the space ship and plugged the thumb drive into the corner of the door. The space ship started to whir and spin and Rick was standing right in the middle of the ship. He was not spinning.

The giants looked around and tried to beam back on their ship and couldn't.

"What have you done?" yelled Bjorg. "Our machines do not beam us up any longer."

Bob smiled and said to Rick over the communicator, "Okay Rick, take her up."

The space ship lifted into the air and the humming stopped. It looked and sounded as if it was floating.

Rich said over the communicator, "This is way too cool. I want to try some maneuvers."

"Go for it," responded Bob.

The giants and the crew with Bob and John watched the ship fly to the North, South East and West and turn quickly at a hundred and eighty degrees.

When Rick landed the giants were all over him. They wanted to see how the ship functioned and what the equations were in the program Bob had worked on.

"With these equations in this program we actually can redistribute the molecules of an object and make it weightless in our gravity. We would be glad to share it with you folks from Sarduk if you will trade us technologies."

Truhann and Blitz walked over to John Ticker and said, "Your man Bob seems to know what he's talking about. Can we talk to him in private?"

John said, "Sure. Just keep me in the loop after you've finished talking."

Truhann, Blitz and Bjorg escorted Bob back to his office in the main building. They sat down at the table in the conference room and when they were comfortable Truhann started:

"Mr. Truehorn, I don't know how you did it but you saw through our bluff. We will take over your planet if you do not cooperate, but we do not have the facilities to do so right this minute. I have eight hundred ships circling one of your three moons at the present time. They are standing by for orders from me. Two hundred ships are ready to land on each of your continents on Trilla. Heartland, Soulland, Plataland and Hankaland. We picked Soulland because it seems you have the most advanced government programmers. Can we cooperate?"

Bob shuffled his feet and thought about what Zaza had told him.

"If you put in writing your willingness to cooperate and not take over I believe we may be willing to work something out."

"You don't understand," said Bjorg. "Your program has already disrupted the waves that travel from planet to planet. These waves are what makes us evolve. Your disrupting of the wave could have grave consequences on all life."

"We have shown you how our running of the ship and the equations makes it impossible for you to beam back up to your ship while our machine is running. Try beaming back up now."

Truhann lifted his transmitter and said, "Try the transporter beam now. I'm ready. On my mark. Now!"

As soon as he said it, he disappeared. Within a minute he reappeared and said, "Okay so when your ship is running it undoes our science and makes our transporter beams ineffective. We need to look at the program and see if we can figure out why"

"Not so fast," said Bob. "I want a written signed document stating you will not try to land and take over before I share anything with you. We do not have the transporter beam technology that you have and we would like it. Perhaps we can negotiate a trade and a peaceful solution."

"I am not at liberty to negotiate for all the people of Sarduk. Our planet is within range of your little Trilla and our planet is four times the size of your little planet. We do have one government, but I am not the leader. I and Bjorg and Blitz are here only as a scouting party. We will have to take your plan back to our headquarters and discuss it before we can work any kind of negotiation."

"Suits me. Let's let Mr. Ticker in on your plan and then we can discuss what Trilla wants to do as well."

Bjorg opened the door and asked John Ticker to join them. When they had given him all the details Truhann said, "We will return as soon as possible. This could take weeks to see what we on Sarduk want to do. Do not lose faith. We will return."

After they left, John Ticker called a meeting of all government officials and the crew and they discussed their options.

"The first thing we need to do is to get the governments of Heatherland, Plataland and Hankaland to join us here in Soulland," said John. "Rick please make contact with them and we will set up a meeting for day after tomorrow at ten in the morning here."

"Will do," said Rick.

The meeting adjourned and Bob got on the phone to Zaza.

"Thanks for your help. You've got to be involved in this process. A meeting has been set up for ten in the morning day after tomorrow. Can you be there?"

"I wouldn't miss it for the world," she cooed into the phone. "See you then."

CHAPTER SIX
'What to do, what to do?"

The time for the meeting came too quickly. . A group of six to eight people from each country was present. Each country had a major emissary. As it happened, Bill Grateful was the emissary from Heatherland. Plataland's emissary was Lacy Smith; and Hankaland's contact person was Smern Golden from Bob's past experience at the International Trust Company and Lee enterprises. They were sitting at a large round table with a stage at one side of it and Bill was at the podium on the stage. Windows around the conference room let them look out onto the surrounding area and the room had an air of openness about it. From the windows you could see a river and rolling hills covered with pine trees. The sky was a beautiful blue color

Bill spoke first and said, "Looking at you Bob, it seems like old home week. Here's Lacy representing Plataland, Smern representing Hankaland and myself representing Heatherland. Because we've had money we all got involved in running things in our country. So, what's this all about?"

John Ticker spoke up and said, "Bob has been instrumental in creating a program that is changing the way the world and the entire universe works. Giant creatures from another planet have visited us here in Soulland and want us to negotiate for the entire planet. We didn't think we could without having representatives from each of our countries here. Welcome to Soulland. We will be briefing you on what is going on for the next hour. Then we will take a break. At eleven we will meet for one big giant brainstorming session. Suzy Lundquist has a firm that will be serving lunch to all of us. I believe most of you know Suzy from your past experience at Lee enterprises. This afternoon we will meet separately to brainstorm and then the plan is to get together at two this afternoon for hashing out what to do. Any questions about the agenda?"

He then wrote the agenda on the board and introduced Rick Burgness as the person that flew the spaceship R12D2.

"When we load the program Bob wrote into the R12D2 space ship and press start the machine begins to whirl and spin on the outside circumference only. It creates its own artificial gravity and by so doing seems to interfere with wave patterns we are just beginning to learn about."

"It seems the Sarduks are using these wave patterns to control their transporters and other items in their lives. When we started the space ship their transporters quit working here on Trilla, at least. We don't know how far into space the wave patterns get generated from our R12D2 but we do know they get generated and interfere with some other kind of wave."

"I'd like Bob Truehorn to come up now and explain what he knows about the equations in the program."

"Ladies and Gentleman," Bob addressed the crowd. "R12D2 is our twelfth attempt at antigravity. It is our second fully functional prototype, hence the name R12D2. It was researched twelve times and built twice. I came upon the equations while watching the waves at my cabin on Lake Fanna, believe it or not. The lake is always a place I go to relax and unwind. Many of my most creative moments come from fishing and relaxing high in the hills on Mount Fanna."

"These equations relate to the early equations that talk about the rate of change, but more importantly they explain how to make things change. Molecules are what we call the smallest parts of everything around us. When we compare living molecules with items that are not living (to us); there appears to be no difference at the smallest level. In other words we are all part of the energy around us and the waves the Sarduks use are no different. Our first ten prototypes used in research did nothing but create energy and cause disturbances on anything electric around us. Now we have developed the R12D2 and all of a sudden we have contact from off planet. People from the planet Sarduk, which revolves around our sun every thirty six hundred years and is approximately four times larger than our little Trilla, are interested in our experiments. We are also interested in how they are using waves to transport people from one place to another and I believe they are using this technology for their spaceships. By fluke our spaceship is more advanced than anything they have. Still we do not understand their technology."

"We have asked them to share technologies with us but originally they wanted to subdue us and use us as slaves for mining operations on different planets. I have invited my friend Zaza Plinkton to our meeting as she was instrumental in our working out an agreement with the Sarduks. Zaza is psychic. I don't know if all of you believe in metaphysical things but Zaza can read the Akashic records and has a few things to say about these people from Sarduk. Ladies and Gentleman, please help me welcome Zaza Plinkton, a marriage counselor and unbeknown to her clients, a very psychic lady."

Zaza took the stage and said, "Thank you Bob. Bob called me on the phone a couple of days ago and asked for help. It seems he felt like these giants were about to take over. I suggested to him that he be very aggressive and not let them push him and the others around as that is how they wanted you to react; Afraid and intimidated. Bob tells me he was aggressive and because of that they are meeting right now to decide if they can do business with us on an equal basis or if they still want to rule over us. There is not much more to tell unless there are some questions."

Bob stood up and said, "We need to decide what plan of action we will take and probably also have a couple of back up plans as contingencies."

Suzy raised her hand and said, "What is it these people want exactly? Do you know Zaza?"

Zaza replied, "Yes I do know. The three that were leading the scouting party wanted to intimidate us and by doing so wanted to feel more self-confidence. Their leaders sent them to get the equations Bob has uncovered or dreamed up or whatever vernacular you programmers use; and so far they have not been successful because Bob stopped them."

She smiled at Bob and batted her eyes again. It made Bob start thinking sexy thoughts and then he remembered she could read his mind so he spoke quickly.

"Zaza visited the Akashic records of the three giants and found out they were trying to build up their egos. I would never have guessed that. Zaza helped save our planet because if I hadn't acted aggressively and if Rick hadn't got to the space ship we may all have been working as slaves in one of their mines right now. Thanks Zaza for your insights."

Everyone applauded and Zaza sat down. John Ticker said, "Let's take a fifteen minute break and when we return we will do some brainstorming about what to do."

Zaza came up to Bob during the break and said, "You shouldn't be thinking those sexy thoughts about me, Big Boy."

"I can't help it Zaza. I am only a man."

"Okay. I understand," she said. "Thanks for saving our planet, this time."

John interrupted and said, "I hope we can work something out that's agreeable to everyone."

"Me too," said Bob and they returned to the conference room.

Bob led the brainstorming session.

"Okay the rules are
1. No one comment about anything that is said either negative or positive.
2. We write all ideas on the board or if need be create more boards to write on. We have these large Easel pads we can write on so we won't run out of space.
3. I have asked Lacy to write down every idea.
4. They do not have to be in order

So if you're ready let's begin.

Bill said, "Arm ourselves and defeat the Sarduks when they try to take over." Lacy wrote it down

Smern spoke and said, "Get their promises in writing backed up by at least three signatures from their government for total cooperation."

Suzy Lundquist said, "Feed them and make them our friends before we do anything more."

Lacy said, "I think we should trade only one thing at a time. For example they give us the technique and equations for transporting and we give them the wave generator. Then we could go on from there for differing technologies."

She wrote all these things on the board as people said them. For her saying she wrote '1 item traded at a time.'

Bill Grateful said, "I think we need to know why they need slaves to mine materials. We need to find out more about their needs and by doing so maybe can help them find better alternatives, such as robots or machines."

John Ticker spoke up and said, "We need to give them a united front. We have four governments represented here and need to make it clear to the Sarduks we are together on this and we are in agreement with whatever we come up with."

Many other things were written down and soon everyone went to lunch. Suzy's catering was phenomenal. She served hamburgers and salad along with baby green asparagus and a vegan burger that looked like real bison. On Trilla the main meat dish was Buffalo or Bison. Everyone enjoyed their lunch and returned to the meeting charged up and ready to hammer out what to do.

John Ticker took control and said, 'We have some wonderful ideas here. Now let's break into groups of four to six and each of us come up with plans we would like to implement. Try to mix the groups up so there is one or two people from each country. Then we will meet again and each group will present their plan. This will probably take most of the rest of the day. If your group needs more time let me know and I will come around and see how everyone is doing. If I figured correctly we should have about eight groups."

Bob's group consisted of Zaza, Bill, two people from Hankaland one person from Plataland. A total of six people. Zaza and Bob were from Soulland, although they both had lived in Heatherland. The two from Hankaland were Rhonda and Larry Richardson, a couple. The person from Plataland was Donna Blisted.

The group decided to have Bob write down their ideas and Donna was to be their spokesman when they met in a larger group. After hashing things around for the better part of three hours this is what they came up with:

>   1.   Invite the Sardukians to a feast celebrating the coming together of their planet. Put out formal invitations and have representatives from all the governments in attendance.

2. Find out why they need slaves to perform mining operations. Help them come up with alternatives.

3. Prepare the military to shoot down space ships and fight against giants as a backup plan if needed.

4. Suggest they start by trading their transporting abilities with one of the six equations necessary for the wave generator that ran R12D2. Then trade technologies on a one to one basis until all six equations are given to the Sarduks. We would like to learn from the Sarduks the following:

    1. how to make our ships invisible,

    2. transport,

    3. where the wormholes are for space travel,

    4. how our equations will allow us to fly faster than the speed of light,

    5. what they have to do with mining,

    6. and help us evolve so we can use more of our brain power than we are currently using.

Surprisingly when everyone got together and made the presentations they were pretty much in agreement. Bob's group had spelled out what we wanted to learn from the Sarduks and Bill's group had spelled out how to prepare the military to fight against giants and shoot down space ships that can become invisible and travel at fantastic speeds. Suzy's group spelled out what kinds of foods and celebration they were going to have, but pretty much all groups agreed on the same things.

Invitations were sent out to Sarduk and all over Trilla for the coming together feast. Bob, John, Suzy and Zaza were going to represent Soulland. Bill Grateful and his wife Doreen, were going to represent Heatherland, Lacy Smith and her president, Victor Greenoble were to represent Plataland, and Smern Golden with his president Andrea Smitten were to represent Hankaland. Rick picked them up in R12D2 and flew them to the park on Dana Bloort's island off the coast of Hankaland where the festivities were to occur. The world was invited. All Trilla people and all of the people from Sarduk were sent invitations. Cost was to be one hundred zilphers and attendance was limited. Bill and Smern contacted Dana Bloort who had purchased his own island, and arranged to have the party on the island.

Suzy had her crew working with Dana Bloort a few weeks before the event to make sure all kinds of food was available and chairs and tables would accommodate sixteen foot tall people.

Dana was as excited as a kid with a new toy.

"I haven't had this much fun since I was a kid," he said as he stroked his handlebar mustache.

CHAPTER SEVEN
'The Party'

Bob was sitting at poolside talking to Dana about how he liked living on his own island when Hanta and Georgina and Ruth arrived together. They had flown over in their little airplane and landed on the ocean and parked at the Marina of the island.

Bob was glad to see Ruth and to see that Hanta and Georgina were still alive and well. Soon they were talking about all kinds of things and never did mention the Sarduks. It was still early in the day. The party was set for five in the afternoon and it was only two.

"I'm going for a swim," said Georgina. "Does anyone want to join me?"

"That sound's like fun," said Ruth and soon the four of them were splashing and playing in the pool.

"I wish everyone could enjoy life like this," Ruth said. "This is so fun."

"It would be nice if everyone did as well as we're doing at Manton," said Hanta. "Hey! Did you hear? We're going to become an employee owned company. Our chief executive officer Dan Manton has been reading about another corporation that did it and they are doing very well. We're having a meeting next week to discuss it. What does that mean? It probably means a raise in pay for me and for Ruth, Eh Ruth?"

Ruth nodded her head and dove under the water and swam around a little.

"That sounds neat," Bob said. "I had a difficult time making it in business, and it probably wouldn't have been difficult if I'd had others helping me run the place. That's why I chose to work for the government. It's been two years now since I started with the government. Wow. I'm already twenty eight years old."

"Me too," said Hanta. "It's hard to believe we're almost thirty already."

"Yes it is," Bob replied. "I want to talk about these waves that the Sarduks refer to. I'm intrigued by them and what they are all about."

"You might get more out of talking with Lacy Smith," Ruth suggested. "She's setting over there by the bar looking lonesome. Why don't you go talk to her before the Sarduks get here?"

"Hey, that's a great idea," said Bob and dove under the water and swam around a little pulling himself out of the water just below where Lacy was sitting. He looked over at Hanta and Gloria and Ruth as they got out of the water on the other side of the pool and Hanta gave him a thumbs up sign.

"Hi, Lacy," Bob began. "I've been thinking about those waves the Sarduks talk about and I'd like to talk to you about them. Do you want to talk with me?"

"I'd love to if you'll buy me a drink," Lacy said removing her glasses and trying to look feminine.

Bob thought to himself, "She's not a bad looking woman without her glasses here by the pool in a swimsuit."

"As I understand the waves the Sarduks are talking about," began Lacy. "These waves have been around forever and are forms of energy we can use. Back in the last century we had a man that tried to get us to use the energy. His name was Bikola Nestla. You've probably heard of the Nestla coil? I'll have a scotch on the rocks, by the way."

"Scotch on the rocks, coming up," Bob said and went to the bar to get her a scotch and him a rum and coke.

He took her the drink and said, "My equations utilize the energy waves to transform items at the molecular level. The program I've written bombards the cells of the structure until things start happening. We can program what we want to happen. It took us eleven attempts with our R12D2 prototype before we figured out how to make it fly."

"Yes. Since I've been involved with research most of my life I can tell you there is a lot of power on those waves you are using. That's why the Sarduks are so interested. I don't know how you came up with the equations by looking at waves on the water, but I would be interested in hearing what you have to say."

"I was fishing and a large Beaver saw me and jumped into the lake where I live and slapped its tail. The waves looked like energy and I suddenly started reviewing the calculus I'd learned in college about the rate of change and the next thing I know I was putting the equations in my laptop back in my cottage. I could almost see the water changing as the waves went along and I was thinking of the fish and how they reacted to the waves and how everything is interconnected."

"Wow. That's quite a story," she said. "I'll have to come fishing with you sometime."

"I'd like that," said Bob. "I have an extra pole anytime you want to come up. In fact my cabin has an extra bedroom so we could arrange it anytime."

"I'd like that," said Lacy. "Victor Greenoble, our President in Plataland, has a wife and family and I don't go out much with people my own age. I guess that's why I stayed with International Trust Company. They have paid me well to do research and I have enjoyed doing something with my life."

"Well you wrote things down very professionally when we were meeting together," Bob volunteered.

She almost blushed and said, "Thank you. Thanks for the drink. It's nice talking to you. Don't you have a girl friend over there?"

"Well she is my friend, but no closer than you are. She helped me run my music store when I tried that out a few years ago. We're just friends."

"Oh, that's nice."

Lacy looked pleased that Bob said Ruth and her were just friends.

"After we get through this meeting with the Sarduks, I have a free weekend coming up weekend after next. Can I invite myself to your cabin for a lesson in fishing?"

"Next weekend it is," Bob replied. He finished his drink and said, "Would you like another drink, my lady?"

"Oh! Such elegance. Of course."

He went to the bar and got refills of the drinks. When he returned Victor Greenoble and his wife Samantha were setting beside Lacy. Their eight year old daughter was getting ready to jump in the pool.

"Have you met Bob Truehorn President Greenoble," Lacy asked?

"No, I haven't had the pleasure except for at the meeting. That was too formal to get acquainted. It's nice to meet you. This is my wife, Samantha and our daughter Lucy. Now Lucy, stay close to the edge until you get better at swimming."

"Bob and I were just discussing the waves of the future and how these waves may affect everything we do," said Lacy matter-of-factly.

Samantha looked at Bob and said, "Come on Vic let's get wet."

They jumped in the pool and Lacy laughed. "It's good to see them having fun. I hope it keeps up for awhile. This could be a very un-fun meeting tonight."

"I've met three of the Sarduks up close and they aren't as scary as they first appear. Sure they are sixteen feet tall but they are people just like you and me," Bob suggested.

"That's good to hear," Lacy said smiling.

"Tell me, what do they say about the waves?"

"Well, one of the giants named Bjorg told me I didn't understand. He said my program has already disrupted the waves that travel from planet to planet. These waves are what make us evolve. Our disrupting of the wave could have grave consequences on all life. That got me to thinking about it. He said it could have grave consequences on all life? I don't know? That's probably why I wanted to talk about it. We need to do some more research on these Waves."

"As it turns out, I've just given my notice to ITC. They're running out of research money and I've decided to start my own company of researchers. I want to start a company where everyone is an owner."

"Hey. That sounds familiar. My friend Hanta works for Manton guitars and they are talking about doing just that; making their company employee owned. I'd like you to meet them. They're great people. Ruth works for Manton guitars too. Come on over and I'll introduce you."

They stood up and walked to the other side of the pool.

Bob began with, "Hanta and Georgina and Ruth, I'd like you to meet Lacy Smith. She was in charge of research for Lee Enterprises and then International Trust Company. She's interested in starting a research company where everyone is part owner. I told her about Manton and how you guys are going to restructure your company like that."

They shook hands around and Ruth looked a little sad. "I wish I could read peoples thoughts like Zaza does," Bob thought.

Hanta said, "Manton is having a meeting next week and we'll be discussing how we are going to make it happen. If you'd like; we could all meet together and discuss it after next week."

"That would be neat," said Lacy.

"Hey! I've got an idea," said Bob. "Let's all meet at my place weekend after next. Lacy want's to go fishing and we can all meet for the entire weekend."

"I can't weekend after next," said Ruth. My sister is getting married over in Hankaland. Smern and Bill have invited me to golf with them before the wedding. You guys go. It sounds like fun."

"Okay let's plan on it," said Hanta. Georgina smiled and nodded her head.

"Buy me a drink hon," Georgina said to Hanta. "I'd like a scotch on the rocks."

"Nice choice," said Lacy as she held up her scotch on the rocks.

It wasn't long and the sky filled with space ships from Sarduk. Everyone was very quiet as the ships landed and people came from them. There were only about six ships. One distinguished looking older gentleman was walking next to a tall beautiful woman. Next to them Bob recognized Truhann and Blitz.

John Ticker greeted them and introduced them to Dana Bloort. Bob ran up to them and said, "Hi Truhann and Blitz. Where's Bjorg?"

Truhann frowned and said, "Bjorg had to attend to some other duties and couldn't make the party. He told us to send his apologies that he couldn't come. This is my ambassador and president of the land of Sarduk, President Jose Guevara and his lovely wife Eve. President and Eve this is Bob Truehorn, the inventor and programmer we've been discussing."

"I've heard a lot about you," said President Guevara.

"Come, come," said John Ticker. "The food and music is waiting."

"Please make yourselves at home on my little home here on the island," said Dana Bloort. "If you need anything do not hesitate to let me know and I will see to it that you're wish is granted. Please, relax and enjoy yourselves."

About fifty sixteen foot tall people from Sarduk walked slowly to the tables and started to intermingle with the little six foot or so tall people of Trilla.

Everyone enjoyed themselves and the evening was a huge success.

"Look at those tall females," Bob said out loud to Hanta. "They must be taller than three of me and you."

"Take me to your ladder, I'll see your leader later," said Hanta trying to be funny.

"I'll say," said Bob. "Wow!"

"This was a good idea setting the tables up so high. We climb the stairs and get to the tables from the back and they just help themselves at the front. Who's idea was that anyway?" asked Georgina.

"I have to give credit to Suzy for that idea," said Bob. "Suzy and Dana have been planning this for over a week. It's turning out nice, so far, eh?"

"Aye Maloha," said Ruth and Lacy frowned.

"Boy, I wish I could read minds like Zaza," thought Bob again as the two women looked at each other funny.

Just then he saw Zaza entering behind the bar.

"Excuse me folks. I've got to say hi to Zaza," Bob said.

He almost ran to her and said, "I was just thinking about you and how I wished I had your abilities."

"I don't think you'd like it if you did," Zaza said matter-of-factly. "Ruth is thinking I wish Bob was different. He's so introverted and Lacy is thinking I wish I was better looking so Bob would like me as well as he likes Ruth. There. Are you happy now. If you want to read someone's mind just come to your local psychic and ask her what someone is thinking and she will tell all and bare all. Buy me a soda pop will you?"

"Yes my lady," Bob said and went to the bar for two soda pops. "I've had enough to drink for the night," he thought.

"Yes you have," said Zaza as he brought her the soda. "Introduce me to your friends. I know Lacy but not the others."

Bob introduced Zaza as a family counselor to Hanta, Georgina and Ruth.

Truhann joined them and over towered them until he set down in one of the seats they had made special for the Sarduks. When he sat down he was at eye level with everyone standing.

"This is a very nice party," Truhann said. "I think it will help in our becoming better acquainted."

Zaza stood up and said, "Excuse me. I have to visit the ladies room." She left and Lacy went with her.

When they were away from earshot of Truhann, Zaza said, "All the time Truhann was talking to us about how nice it was; he was looking around to see how easy it would be to make us his slaves. That's what he was thinking and our emissaries need to know that."

"Yes, they do," said Lacy. "Are you sure?"

"Yes, I'm very psychic and can read others thoughts. Sometimes it gets me in trouble," said Zaza.

"I can imagine," said Lacy. "Have you ever considered working in research? I'm starting a research company in a couple of weeks and it would be very helpful to have someone with your talents helping us."

"That sounds like it could work. My counseling is slow right now. I cure people too quickly because I can see what their real problems are."

The two ladies returned from the ladies room and joined the party.

"Let's have fun tonight anyway," suggested Lacy.

"Good idea," said Zaza and they went to the bar and got their own drinks; Lacy with Scotch on the rocks and Zaza with soda pop.

When they got back together with Bob, Hanta, Ruth, Georgina, and Truhann; Blitz had joined them. He was sitting across from Truhann.

Zaza said, "Why do you need slaves and why are you thinking of using us as your slaves?"

John Ticker said, "You'll have to excuse her. She just blurts out whatever is on her mind."

"Well," said Blitz. "We are too large to go down into the mines that have been drilled. We have mines for many different kinds of ores and minerals we need for our way of life. The larger we have to drill the holes the more money it takes to remove the minerals. Yes, we have been thinking of using you folks with littler bodies to mine for us; to save money." He grinned from ear to ear and Hanta spoke up:

"It seems to me you should do what we do. We have mines all over Soulland but they are mined by robots. Small little things that dig and chop and don't get sick from bad air and can't die."

"You really have that technology?" asked Truhann.

"Yes, we do," said Hanta.

"Perhaps we can make a trade. We have never thought about robots. Here we thought we were more advanced than you folks because we're flying all over the galaxy and you're not."

"I'd like to know where the wormholes are so we can travel quickly from one solar system to another. Or even within the solar system to save time," said Bob looking at Blitz. "I'll tell you what; we'll help you build some robots for your mines if you'll show us where the wormholes are."

"Just a minute, just a minute," said John Ticker. "We've put together on a list all the things we'd like to know from you and what we would like to trade for them. You guys need to put together a list also. But for now, let's enjoy ourselves and save this trading talk for the office time. Have you gentlemen tried our Trilla beer? We have a brewery right here in Bloort Island and Dana is famous for distributing it all over Hankaland."

"You're right," said Blitz. "Today is party time. Let's see if we can enjoy each others company. Let me try the Trilla beer."

"I'll be right back," said John and he went to the bar.

While he was gone, Zaza said, "I can read thoughts so you don't have to say anything out loud. For example, Mr. Truhann I know you were thinking how easy it would be to overcome us and make us your slaves a little earlier. I don't think you will find it as easy as you think."

Truhann replied, "You really can read thoughts? I'd heard of that happening but never met anyone that could do it. Let's try a little experiment, tell me what I'm thinking?"

Zaza said, "I see a queen of spades in your minds eye. Tell us, is that what you were concentrating on?"

"Why yes it was. This is very entertaining. Let me try some more. Okay I'm concentrating on something. What is it?"

"You're thinking about the Trilla beer we have just been told about. Would you like a drink? John brought back enough of them from the bar so we could all try it."

"Yes, I would," said Truhann. "This is a fun party."

They went back and forth a while and finally Zaza said, "I'm going swimming. Have you tried our pools? The one pool is over thirty feet deep so you will have to swim."

"Let's go swimming," said Blitz and the entire party jumped into the pool and swam around for a while.

CHAPTER EIGHT
'Robots for the mines'

The next week the emissaries met with the President of Sarduk and showed him a list of things they wanted and what we would trade in return.

"Let's get started," said President Guevara. The first thing on their list was trading the robots for a look at where the wormholes were in the galaxy. John and Bob were surprised to see two of them on the back sides of two of Trilla's moons. Soulland contracted with ITC to build robots for Sarduk and they traded for the maps showing where the entryways were in the galaxy. Until they could fly the new R12D2 they couldn't check out most of the passages so they had to take the word of the Sarduks.

President Guevara arranged for Bob and John to fly with him to four of the wormholes on their ships. After doing so, they traded one hundred of the little robots for the information.

"It occurred to me, we could use the technology you use to fly our space ships and then wouldn't have to disrupt the waves of the future as you say our new ship is doing." Bob explained to the President that we should probably add this to the list.

"I'll get Bill Grateful to go with us to see how you build your spaceships and I'll give you the first equation of the six needed to alter the wave theory," said Bob.

"Let's make it happen," said President Guevara.

They continued on with exchanges and trading very peacefully until Bob wanted to know more about the waves.

It had been a little over a week since the party and Trilla had received:
1. Space Age technology to fly ships across the galaxies.
2. Locations of wormholes or places where a space ship could travel instantly large distances.
3. The ability to make our ships invisible
4. How to transport beings from one location to another.
5. Flight faster than the speed of light

The Sarduks had received:
1. Equation one of R12D2
2. One hundred robots
3. Equation two of R12D2
4. Equation three of R12D2
5. Equation four of R12D2
6. Equation five of R12D2

They were stuck on the last item on their list. Help us evolve so we can use more of our brain power than we are currently using.

President Guevara addressed the assembly of people in a meeting one and a half weeks after the party. Everyone that was at the party was there. He said, "The last equation will give you the means to affect the waves of the future. It is these waves that your equation is affecting and it is these waves that allow us to transport items from one place to another. We have shown you how to transport items from one place to another and also how to fly faster than the speed of light. Your equations and your program nullify our being able to do that. Our best scientists cannot figure this out. We cannot help you until you help us. Get with your best scientists and we will meet in two weeks to see if anyone has made any headway. You have the advantage as you have the final equation and we do not. I think our scientists are clever however so I am putting all faith in the outcome of the next two weeks. I'll see you then."

The following Friday evening, Hanta and Georgina picked up Lacy and flew her to lake Fanna where Bob was waiting for them to show Lacy how to fish and Hanta and Georgina how to relax.

"Welcome, welcome," Bob said as the threesome landed on the water and taxied to the marina next to Bob's cabin.

"This is beautiful," said Lacy as she looked around. I can't wait for you to take me fishing."

"First thing in the morning," Bob said. "Let me show you where you will spend the night."

He proudly showed Lacy the third bedroom reserved for special guests. Hanta and Georgina had already settled in the second bedroom of the cabin.

They went outside and the four of them watched the three moons come up. about two hours apart. Two of them were already half way and two thirds the way across the sky when they went outside.

"I could learn to like it here," said Lacy dreamingly.

"I love it here," said Bob.

"We do too," said Hanta and Georgina in unison. They smiled when they said it together and everyone laughed.

"What a perfect night," thought Bob and Lacy said, "It's perfect, isn't it."

"Oh no," thought Bob. "Lacy can read minds too." He looked distressed but Lacy didn't react so he thought to himself, "Am I losing my mind?"

Soon everyone yawned and turned in for the night.

CHAPTER NINE
'The Company You Keep'

Six in the morning and Bob was making coffee. Lacy entered the room and she was showered and dressed.

"I helped myself to taking a shower. This is a lovely home."

"I'm glad you helped yourself. I meant it when I told you to make yourself at home. Would you like some coffee?"

"I'd love some."

Bob was thinking how good she looked without her glasses and no makeup. She had a nice body and he couldn't take his eyes away from her.

"So you're going to show me how to catch fish here in Lake Fanna? I must warn you, I was quite a tomboy growing up and used to out-fish my father."

"Is that so? Catching fish is a little different here in Lake Fanna. You have to be able to feel the fish nibbling and hook them just at the right time. It's really quite relaxing."

"So are we going out before the others wake up and before breakfast?"

"Why not? If you're ready let's go."

Bob picked up the two fishing poles and the tackle box and they walked down to the marina. It was a beautiful morning. The sun's rays were lighting everything up even though it was not visible yet. In the distance they could here the sound of a Loon.

"Listen to that Loon," Lacy exclaimed. "Isn't this beautiful?"

They got in the boat and Bob took her over to the side of the lake where the spring bubbled up. They could see trout swimming around under the boat. Bob stopped the motor and said, "This is where you'll catch some fish."

They both started fishing and soon Bob had caught three fish.

"I can feel them nibbling but I can't seem to catch them," Lacy said.

"Here, let me show you," said Bob as he reached around her and held the pole with her. Her smell made Bob excited as he held her in his arms from the back.

"Do you feel that?"

"Yes, I can feel it. It feels like they are touching the line."

"Well, the secret is to jerk just a little when you feel it on about the third tap. Like this." He jerked a little on the line and her pole bent with a fish on.

When she had pulled in the fish, she said, "That was fun. I think I liked you showing me how more than I liked catching the fish."

Bob felt himself blush a little as she removed the fish and put it on the fish line. She cast her line out again and before Bob could catch another fish, Lacy had caught four nice trout.

"Hey! I've got one more than you. I warned you," she said smiling.

When they had caught about six fish each Bob said, "Let's go cook these for breakfast for Hanta and Georgina."

"Okay," said Lacy and they pulled in their lines and went back to the cabin. After they had cleaned their fish, Lacy mixed up some flour and started the cooking while Bob prepared some eggs and more coffee.

Hanta and Georgina entered the kitchen and Georgina said, "Something smells very good. Is that trout I can smell?"

"Freshly caught," said Lacy.

"Lacy is quite a fisher woman," said Bob to Hanta and Georgina. She caught four fish before I could catch one."

"Yes," said Lacy. "But Bob had caught three and had to show me how before I did."

They smiled at each other and Hanta and Georgina smiled at each other.

"This is delicious," said Hanta as they were eating their trout breakfast. "Maybe I should take up fishing too."

"I have four poles and there is room on the boat anytime you guys want to go. You can even take the boat out yourselves if you want; anytime."

"Manton enterprises have offered to make all employees owners of the company. They gave all of us a copy of the agreement they want us to sign to look at over the weekend. I brought it with me because I knew you were interested, Lacy."

"I am," Lacy replied. "I want to start my own employee owned company. I believe it's the only way to have a company in this day and age."

"After breakfast I'll get it for you," Hanta said.

"That will be great."

Bob said, "If companies become democratic and have better wages for everyone, I might reconsider starting a business again. My job is fun and important, but we still give many of our contracts out to private industry. Too bad I didn't make my music store an employee owned company."

"If you had, it would probably still be around," Hanta said. "It's hard making it in a small business when you are the only one that cares about the outcome."

"Ruth was great as a salesclerk and helper, and I gave her a living wage. The sales weren't enough to make it."

"I heard you need enough money to last for several years at not making money if you're going to make it in business," Lacy volunteered.

"I've heard that as well," said Georgina.

"Our sales at Manton have been very high. Ruth is a great sales lady," said Hanta.

"I'm glad things are working out," said Bob. "Do you guys want to go for a hike after breakfast?"

"Sure," they all said in unison and then they looked at each other and everyone laughed.

After breakfast Hanta brought a copy of the agreement for being a company owned employee from Manton to Lacy and they all went hiking.

They hiked around Lake Fanna and saw the beavers, several deer and a Mother and baby Moose.

"Don't get too close to the baby Moose or the mother will charge you," warned Hanta.

Lacy wanted to pet it, but decided against it when Hanta said that.

"Such a pretty little thing," Lacy said.

"It is pretty," said Georgina.

"You won't even think about that if the mother charges us," Hanta offered.

"Come on, I want to show you a waterfall up over the next ridge," said Bob and they continued going up the trail over the next ridge.

The waterfall was over fifty feet high and the air was wonderful right below the falls.

"I'm going for a swim," Lacy said and she took off all her clothes and dove into the water. Everyone followed.

"Brrr," said Hanta. "I'm not staying in, but it is refreshing."

"You can say that again," said Georgina.

"I'm not staying in, but it is refreshing," teased Hanta.

"You guys are too much," said Bob as he climbed out of the water and put his clothes back on. "Thanks Lacy for daring us to jump in without saying a word."

"You're welcome," she said. "You guys are fun."

They all put their clothes back on and hiked down the trail back to the cabin.

At the cabin they started to discuss different kinds of employee owned businesses.

Hanta said, "Some employee owned businesses only offer employees a share of the profits or a stock option. They still pay these employees a small wage and make up for it with stock options."

Lacy said, "Yes and some companies pay everyone a living wage. The Chief Executive Officer doesn't make a lot more than the janitor or clerk."

Georgina said, "I read the other day the average CEO is making over four hundred times what the lowest paid employee is making."

"That's not good," said Don. "I wouldn't want to work for a company like that."

"My business is going to give everyone a livable wage and everyone, including the janitor will have a say so in how the business is ran," Lacy exclaimed.

"Sounds too good to be true," said Hanta.

"Yeah. Do you think you can still make a profit?" asked Georgina.

"If I work it right," Lacy said.

The weekend ended and everyone went back to work except Lacy. She got together with Zaza and started to work out the details about the company they wanted to create.

Meanwhile Hanta learned from Ruth that she got engaged to Smern Golden over the weekend.

"We're going to be wed in two months," she told Hanta. "I spent the weekend on his ranch in Hankaland with Bill Grateful and his family. They all play guitars and they had quite a jam session. I think I sold another four Mantons this past weekend."

"That's great," said Hanta. "I can't wait to tell Bob."

"Hey. Let me tell him. Remember, I worked for him for several months and he is a great guy."

"You've got it," said Hanta.

"How did your weekend with Lacy and Bob go?" she asked Hanta.

"Oh, we had a lot of fun. Lacy caught as many fish as Bob and we hiked all over around the Lake."

"That sounds like you guys had fun. We did too. Ain't life grand?"

Hanta and Ruth went back to their respective offices.

Zaza and Lacy were having lunch and discussing the company they wanted to form.

"I think we should have 'Warriors' in the title of our company," said Lacy to Zaza. Bob gave me a copy of Don Miguel Ruiz's book 'The Four Agreements' over the weekend and I like the idea of being a warrior."

"That sounds good to me," said Zaza. "Has Bob heard anymore from the Sarduks?"

"Apparently they are hung up on the last equation trade. Seems like there are waves that have something to do with our evolving and it relates to the series of equations that change matter and stop some kinds of things from happening," offered Lacy.

"This sounds like a job for Research Warriors," said Zaza.

"Hey. I like the sound of that. Research Warriors. Let's make that the name of our company."

"I'd like that," said Zaza. "You know you can probably get a government contract since we are females and they call us a minority still."

"Hey that's right. I wonder if we could convince Bob, John Ticker and the Sarduks that we could help them. Say for about two million dollars. Of course we would have to be given privy to all the Sarduks technology and to the equations and things our government is involved in"

"Of course."

"Let's get started," Lacy said. "Are you in agreement that we will share in the profits fifty percent each and if any other people join us we will rewrite the contract to share in the profits?"

"Why don't we just write it once for an equal share of the profits and then it won't have to be rewritten if someone else joins us," said Zaza.

"I can see you're going to be a big help. Great idea"

The following month the Research Warriors met with the government of Soulland and the representatives of Heatherland, Hankaland, Plataland and the Sarduks to discuss what was needed for the furthering of their getting along with each other.

President Jose Guevara said, "We will not be able to give you the ability to evolve and use more of your brain power, like we do, until we have connected the equations Bob developed and the Waves that enable us to Transport and fly. We do not understand the connection so to us it will be worth the two million dollars if you can help us figure this out."

John Ticker suggested, "I suggest Sarduk pays One million and the four countries here on Trilla pay the other one million. That would be one quarter million each for Hankaland, Heatherland, Plataland and Soulland. How does that sound to everyone?"

"Before we started the Research warriors I turned over the running of Plataland to Richard Smirnoff. Everyone in Plataland agreed with my decision."

Richard Smirnoff said, "Plataland agrees."
Bill Grateful said, "Heatherland agrees."
Suzy Lundquist said, "Soulland agrees and Smern Golden said, "Hankaland agrees."

"Is everyone in agreement then?" asked John Ticker.

"It appears so," said President Guevara from Sarduk.

A contract was written up and Zaza and Lacy started investigating and researching all of the differing technologies and equations.

CHAPTER TEN
'Waves of the Future'

Bob told Lacy several days later while they were eating dinner at Suzy's café, "I believe there is a wave structure of matter in space from which we can obtain the fundamentals of math and physics. All of us experience existing in space. My equations came to me when I was watching waves from the water a beaver had created. Now these equations I have programmed actually upset and control the molecular structure of the flying machine we have created. Why it interfered with transportation of beings or items is another matter."

"I believe with Zaza's help we have found natural signals occurring in our space that affect our brain waves and our evolution. Zaza thinks all biological entities interact with this electromagnetic wave," Lacy offered."

"Did she say affect our evolution?" asked Bob.

"Those were her words. In fact she says this wave is what has made us evolve all along and is why she can visit the Akashic records. She also said this wave makes it possible for her to have extra sensory perception."

"That would explain why the Sarduks didn't want to give us our final request in exchange for the final equation. Our final request was for us to evolve so we could use a higher portion of our brains. I think the Sarduks know more than they are letting on," said Bob.

"They have been a little sneaky about what they tell us and what they don't tell us regarding their technology," said Lacy.

"If my belief is confirmed, which it sounds like Zaza is saying it is; then I understand why all of this works the way it does. This wave is our wave of the future. People will not only be able to use more of their brain power, but we could all become like Zaza and have ESP and tune into each other better."

"I think the secret is in the way the Sarduks transport themselves from one place to another," said Lacy. "The transportation system stops when you fire up your space vehicle. Something in those equations and the interaction of the electromagnetic force is disrupting these waves and causing things to not act naturally."

"This is exciting stuff," said Bob. "I hope you guys get to the bottom of it soon."

"Me too," said Lacy. "On another note, have you seen Ruth and Hanta lately?"

"They are coming over to Lake Fanna this weekend. They said they have some news. Bill Grateful and his wife Doreen and their twelve year old daughter Nicki is coming over with them. In addition Smern Golden is coming with Ruth, Hanta and Georgina. It will be a houseful. Bill said he was bringing his tent and wants to camp out. Will you join us?"

"I'd love too," replied Lacy.

"Do you want to come over tonight?" asked Bob.

"I'd love too," replied Lacy.

That night was a wonderful night for Bob and Lacy. She not only spent the night but when the weekend came she was there as well. Bob thought he was falling in love with Lacy and Lacy thought she was falling in love with Bob.

When everyone arrived, Lacy asked Nicki if she wanted to go fishing. Of course, twelve year old Nicki said "Yes!" enthusiastically so Lacy took Nicki fishing in Bob's boat while Bob helped everyone get situated for the weekend.

In less than an hour Nicki and Lacy brought back enough fresh trout for everyone to have a nice trout breakfast. Smern and Ruth did the cooking and it was generally a cooperative effort. At breakfast Ruth said, "Everyone, we have an announcement. Smern and I are going to get married in less than two months."

Bob and Lacy congratulated them and everyone else already knew about the upcoming wedding.

The weather was perfect that weekend at Lake Fanna. Everyone was able to get in a little hiking, boating and swimming besides enjoying each others company.

Zaza and Lacy went to Sarduk to visit with the engineers and research crew on that planet. When they returned they got very busy and actually hired several more people to help them. Now their office in Plankton was very busy and Lacy decided to spend a little of the millions she had banked since ITC.

"I'm going to have a thirteen story high building built on the corner of Rave and Winslow overlooking the river," she told Zaza as she entered her office. "I'll use my money and it will be my building but I want to name it the Research Warrior building. Do you think everyone will go along with me?"

"I don't know," said Zaza. "Let's find out. "Zaza got on her office intercom and said, "Everyone please come to my office immediately if you can."

Soon the four others and Zaza and Lacy were sitting around the office.

"What is it," said Tammy. "We've just about figured out the Sarduks transporting system. JoAnne, Rachel, and Donna has been helping me and it appears the Sarduks are using an electromagnetic wave that exists in the space above the planets to change the molecular structure of items into pure energy and then back again into whatever they were before starting the system."

"Lacy wants to build a Research Warrior building and she wants us all to go along with her naming it," said Zaza.

"I'm going to use my own money to build it," Lacy said.

Rachel said, "I think having our own building would be great. Why don't we use the profits from our two million dollar contract and then it will be the Research Warriors building?"

"I agree," said Donna the fourth person Zaza and Lacy had hired. "Having our own building will not only help with our advertising in getting us more business, but we can rent out offices to others in the building and have earnings from that as well."

"Okay," said Lacy. "Let me show you the plans. We have thirteen offices on each floor and thirteen floors. Thirteen is our lucky number. The building I had designed by one of the architects on Sarduk. It looks kind of space age. Here is an artist's rendition of it."

She held up the picture for everyone to look at.

"I like it," said Zaza.

"Me too," said the others in unison and then they all looked at each other and smiled.

"I want to thank you Lacy and Zaza for hiring JoAnne and I, after our bank went bankrupt," said Tammy. "JoAnne and I like it better here. Working together to help the world is a lot more exciting."

"It sounds like you're doing some good stuff," said Lacy.

"We hope so," said JoAnne.

"Okay then. Everyone agrees on going ahead with the building. I'd like it to overlook the river. The lot on Rave and Winslow is available," said Lacy. "Is that okay with everyone?"

"Of course," they all said in unison.

"Have you ever noticed how we all say things in unison sometimes?" said Donna.

"Its synchronicity," said Zaza.

"I'd like to have Bob come over and look at the stuff we've found from Sarduk," said Rachel.

"Yes," said Lacy. "He is very familiar with the equations and how the space ship works and maybe he can see something we don't see."

"He is trained in mathematics," said Zaza.

"I'll call him tonight," Lacy said and she smiled at the crew and they all went back to work.

The next morning Bob made arrangements with his boss, John Ticker to go over to visit the Research Warrior office.

"Hi Bob," Zaza said as Bob entered their small office on Winslow. "Come in and I'll call a general meeting."

Zaza got on the office intercom and said, "Bob's here everyone. Come to a meeting in my office."

Soon they were all sitting around the office.

Tammy said, "We've just about figured out the Sarduks transporting system. JoAnne, Rachel, and Donna have been helping me and it appears the Sarduks are using an electromagnetic wave that exists in the space above the planets to change the molecular structure of items into pure energy and then back again into whatever they were before starting their transport system."

"Yes," said Lacy. "We thought you could look at the program they are using and maybe figure out the mathematics of their system."

Bob smiled and said, "It'll cost ya'."

He grinned from ear to ear and the ladies all looked at him funny.

"Just kidding. I'd love to look at what you've found out and help try to figure it out."

Tammy and JoAnne showed him the equations that made the transportation system work. Bob looked puzzled. "I've never seen anything like this. Could I transfer these equations into my laptop and think about them for a while?"

"Sure Bob," said Zaza. "If you come up with anything please let us know immediately."

"You've got it girls," Bob said.

He then connected his laptop to their computer and downloaded the equations and program. He was looking at it as it was downloading and he said out loud, "Hmmm. Very interesting."

Bob returned to his office and the Research Warriors returned to figuring out the other technologies the Sarduks had shown them.

"This method of flight will revolutionize our little Trill," said Lacy to Zaza as they were studying the programs the Sarduks had given them.

"This is good stuff," said Zaza.

Meanwhile Bob was trying to figure out the equations and what they meant for transportation. Soon the weekend came and he found himself alone at Lake

Fanna. It was Saturday morning and he couldn't get the program he had looked at from the Sarduks out of his mind.

"I think I'll go fishing," he thought to himself and he got his pole and tackle box and went to the boat.

It was another beautiful day on Trill. The Sun was just rising over the Eastern horizon and birds were singing. Bob recognized the call of the Loon.
"I miss Lacy," he thought.

Absentmindedly he got in his boat and went to the spot where the beaver lived. Two large splashes interrupted his train of thought and he watched the waves rock his boat.

"Eureka! That's It!" he exclaimed out loud and he put down his pole and hurried back to his cabin.

He looked at the equations and called Lacy on the phone, "Hey hon. I figured something out and wanted to tell you. Can you come over or should I come there?"

"I'll be right over," said Lacy. It only took her two hours to catch an air taxi and get delivered to the cabin at Lake Fanna.

"Okay cutie pie. What have you discovered?" she said as she was welcomed with a big hug from Bob.

This fourth equation in the transportation system sends out waves that work in contrast with the waves generated by the waves of our space ship. If we put a radiant barrier around the space ship we can still fly it and it won't affect the transportation system. I'm sure of it."

"Wow! President Guevara will be happy to hear this. Can it wait until Monday or should we get everyone together now?"

"I think we need to test it by building the radiant barrier and trying to transport something at the same time before we tell anyone. Just in case I'm wrong."

"Okay when shall we do it?"

"Let's get Rick and John and the crew together right now over at the military base."

"Right now?" Lacy cooed.

"Well, maybe in a little while," He gave her a long and hard kiss and took her into his bedroom to play.

CHAPTER ELEVEN
'Propinquity'

"Will you marry me?" Bob asked as they were traveling to the place he worked.

"I'd be very happy to," Lacy said smiling and snuggling up to him.

"Will you come to work with us at Research Warriors," Lacy asked.

"Only if you stay on as the Chief Executive Officer," said Bob. "These government contracts quite often only go out to companies that have women running them."

"The government is comprised of a bunch of sexists," said Lacy.

"Yes, it is," said Bob.

They showed the crew the differing programs and Bob explained them to the crew.

"It seems there is a wave structure of matter in space that gets involved when we start up our program. When we use the program the Sarduks showed us to transport things from one place to another it uses the same wave structure. I have it under good authority this wave structure is responsible for us using more of our brains and for our evolution. When we fire up our space ship it uses the same wave structure. In order to keep things in balance I think we need to create a radiant barrier around our systems that transport things as well as around our flying vehicle. If we do that, we can continue using the wave energy and not have any problems. Lacy and I have designed a radiant barrier we would like to try to see if it works. Are you guys up to it?"

"Let's do it," said Rick and the crew looked at the plans and built the barrier.

When they had it completed they built another one around the transport system John had made from the plans the Sarduks gave them.

"Okay. Start the flying ship Rick," Bob said.

The machine whirred and lifted off the ground only about fifty feet. Rick got on his intercom and said, "I'm hovering fifty feet high."

"Okay John. See if you can transport one of the crew from one place to another."

John pointed the transporting beam and pushed some buttons.

"It worked!" He yelled.

"Okay let's tell the Sardukians," said Bob and Lacy gave him a hug.

John arranged with Dana Bloort for a huge party and meeting with everyone that was at the first party some several months earlier.

Lacy and Bob were excited when President Jose Guevara said, "We will now give you the ability to evolve and use more of your brain power, like we do, because you have connected the equations Bob developed with the Waves that enable us to Transport and fly. We understand we need to put radiant barriers around all our transporters and now we can utilize the flying ships like you do thanks to your giving us the final equation."

Suddenly Bob felt a wave that seemed to make him feel lighter somehow.

The wave swept over the entire party and they all looked at each other and started talking to each other without using words.

"We're getting married," Bob said out loud and pointed to Lacy.

"Congratulations," said Ruth.

www.ingramcontent.com/pod-product-compliance
Lightning Source LLC
Chambersburg PA
CBHW070153230526
45471CB00002B/638